OUR SOLAR SYSTEM

Uranus

BY DANA MEACHEN RAU

Content Adviser: Dr. Stanley P. Jones, Assistant Director, Washington, D.C., Operations, NASA Classroom of the Future
Science Adviser: Terrence E. Young Jr., M.Ed., M.L.S., Jefferson Parish (La.) Public Schools
Reading Adviser: Dr. Linda D. Labbo, Department of Reading Education, College of Education, The University of Georgia

COMPASS POINT BOOKS
MINNEAPOLIS, MINNESOTA

For Russell

Compass Point Books
3722 West 50th Street, #115
Minneapolis, MN 55410

Visit Compass Point Books on the Internet at *www.compasspointbooks.com*
or e-mail your request to *custserv@compasspointbooks.com*

Photographs ©: NASA, cover, 1, 3, 6 (inset), 14–15, 16 (bottom), 20; Bettmann/Corbis, 4, 6–7; North Wind Archives, 5; NASA photo courtesy of Space Images, 8–9, 13; Astronomical Society of the Pacific, 9 (top), 16 (top), 17 (all), 18-19 (all); Courtesy NASA/JPL/Caltech, 10-11 (all); Corbis, 21, 24–25.

Editors: E. Russell Primm and Emily J. Dolbear
Photo Researcher: Svetlana Zhurkina
Photo Selector: Dana Meachen Rau
Designer: The Design Lab
Illustrator: Graphicstock

Library of Congress Cataloging-in-Publication Data

Rau, Dana Meachen, 1971–
  Uranus / by Dana Meachen Rau.
    p. cm. — (Our solar system)
  Summary: Briefly describes the discovery, composition, planetary motion, moons,
and efforts to study the planet
  Uranus. Includes index.
  ISBN 0-7565-0299-3 (hardcover)
  1. Uranus (Planet)—Juvenile literature. [1. Uranus (Planet)] I. Title.
QB387 .R38 2002
523.47—dc21   2002002949

© 2003 by Compass Point Books
All rights reserved. No part of this book may be reproduced without written permission from the publisher. The publisher takes no responsibility for the use of any of the materials or methods described in this book, nor for the products thereof.
Printed in the United States of America.

Table of Contents

- **4**   Looking at Uranus from Earth
- **8**   Looking at the Way Uranus Moves
- **11**   Looking Through Uranus
- **16**   Looking Around Uranus
- **19**   Looking at Uranus from Space
- **23**   Looking to the Future
- **26**   The Solar System
- **28**   A Uranus Flyby
- **28**   Glossary
- **29**   Did You Know?
- **30**   Want to Know More?
- **32**   Index

Looking at Uranus from Earth

What do you do in summer? Do you swim at the beach? Do you eat ice-cream cones? Summer on Earth is about three months long. That may seem like a long time, but summer on Uranus lasts twenty-one years! You would have lots of time for swimming and eating ice-cream cones if you could live on Uranus!

People on Earth have known about the planets Mercury, Venus, Mars, Jupiter, and Saturn for thousands of

Sir William Herschel discovered Uranus in 1781.

years. But Uranus was not found until 1781. An English astronomer named Sir William Herschel located Uranus with his telescope. People had seen Uranus before, but they thought it was a star. Herschel knew it was a planet because of the way it moved in the sky. It moved differently from stars.

Herschel (1738–1822) worked for King George III of England. Herschel wanted to name the planet the "Georgian Planet" after the king. Other people wanted to name the planet "Herschel." A German

◀ Herschel wanted to name Uranus the "Georgian Planet" after King George III of England pictured at left.

astronomer named Johann Bode thought it should be called *Uranus*. All the other planets had been named after Greek or Roman gods. Uranus was the Greek god of the heavens. People finally decided to call the planet Uranus in 1850.

You need **binoculars** or a **telescope** to see Uranus in the sky. Uranus looks like a blue-green ball. Details on its surface are blurred because Uranus is so far away. People knew little about this planet for a long time.

Uranus (inset) is so far away from Earth that it is very hard to see any details on its surface.

Uranus was named after the Greek god of the heavens.

Looking at the Way Uranus Moves

All planets in the **solar system** travel around the Sun, or revolve, in a path called an orbit. Uranus takes eighty-four Earth-years to make its trip around the Sun. All the planets also spin, or rotate, as they revolve. It takes Uranus about seventeen hours to spin around once. This is the length of Uranus's day.

Most planets spin standing up and down, like a top. Uranus is tilted on its side. So its rotation is quite different.

The gas methane makes Uranus appear blue-green in color. The light patches in the photo are clouds on the planet.

It spins on its side, like a ball rolling along the floor. Scientists are not sure why Uranus is so different. They think a large object may have crashed into Uranus a long time ago. That crash may have

▲ *The Sun's rays shine on Uranus as it orbits.*

knocked Uranus on its side.

The tilt of Uranus means the planet has very strange seasons. Each season lasts twenty-one years! In summer, the Sun is out all the time. It doesn't set at night. It is always dark in winter. The sun is never visible in the sky. The sun does rise and set every day in the spring and fall.

What would it be like to live on Uranus in summer and winter? You wouldn't know when it was time to go to bed!

Uranus is the only planet that is tilted on its side.

Looking Through Uranus

Uranus is a gas giant planet. This means that Uranus is made up mainly of gas. Jupiter, Saturn, and Neptune are also gas giants. Uranus does not have a solid surface. A spacecraft could never land on Uranus. Uranus does not have solid ground for it to land on.

Scientists have two ideas about what Uranus looks like inside. Some think Uranus has a small, hard, rocky CORE in its center. A thick layer of

◀ *(From left to right) Neptune, Uranus, Saturn, and Jupiter are all gas giants.*

water, gas, and ice surrounds the core. Next comes Uranus's atmosphere. An atmosphere consists of the gases around a planet. Other scientists think Uranus has no core. They think the whole planet is just a big ball of slush surrounded by a gassy atmosphere.

Uranus's atmosphere is made up mostly of hydrogen and helium. It also has a little methane. This gives Uranus

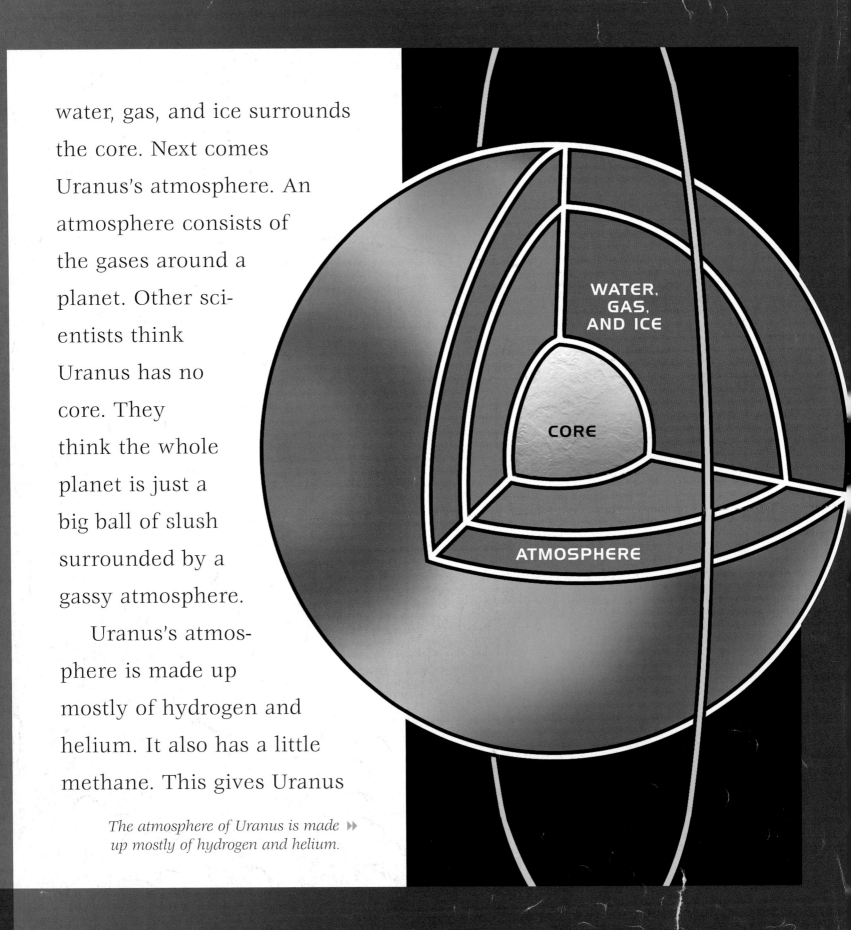

▶▶ The atmosphere of Uranus is made up mostly of hydrogen and helium.

its blue-green color. The atmosphere is also filled with clouds. Winds blow the clouds around the planet from east to west at 90 to 360 miles (145 to 580 kilometers) per hour.

The Hubble Space Telescope is a telescope that orbits in space. It gives scientists a closer view of objects in space than they can get from telescopes on Earth. In February 2000, Hubble saw some very strong storms on Uranus during the planet's springtime. Huge **hurricanes** raged throughout Uranus.

Scientists had not seen

such windstorms before. The last springtime on Uranus was eighty-four years ago, before the Hubble Space Telescope was even made.

Uranus has different seasons. But the **temperature** on Uranus stays about the same all the time. Uranus is far from the Sun. So the Sun doesn't warm up Uranus. The temperature is always very cold, about −320 degrees Fahrenheit (−195 degrees Celsius).

◀ *The Hubble Space Telescope is able to get clearer pictures of Uranus than scientists can get from telescopes on Earth.*

Looking Around Uranus

✦ Herschel did not find only Uranus with his telescope in 1781. He also found two moons orbiting the planet. They were named Titania and Oberon. In 1851, another astronomer, William Lassell, found the moons Ariel and Umbriel. Gerard Kuiper discovered the moon Miranda in 1948.

Titania is Uranus's largest moon. Oberon is the next largest. Oberon's surface is covered with deep holes

Some of Uranus's moons: Titania (top) is the largest. Oberon (bottom) has deep craters.

called **craters**. Umbriel is the third largest moon and is very dark. It has many more and larger craters than Ariel and Titania. Ariel is very bright. Miranda is one of the strangest objects in the solar system. Half of Miranda is ice and the other half is rocky material. Miranda has many **canyons**. Some are 12 miles (20 kilometers) deep. The surface of Miranda looks like mixed-up puzzle pieces. Miranda looks like it fell apart and was glued back together again.

Scientists thought Uranus

◂ *Umbriel (top) is very dark. Miranda (bottom) is covered with strange features that make it look like a puzzle put together the wrong way.*

had only five moons until 1986. Spacecraft and telescopes found many more. Uranus has twenty-one known moons. They are made of ice and rock. There are eleven very small inner moons. Five large moons and many other moons orbit far from the planet.

Uranus also has rings. So do the other gas giant planets. American astronomer James L. Elliot discovered the rings in 1977. Uranus has eleven rings. They are dark and made up of icy boulders. Some pieces are large and some are as small as dust.

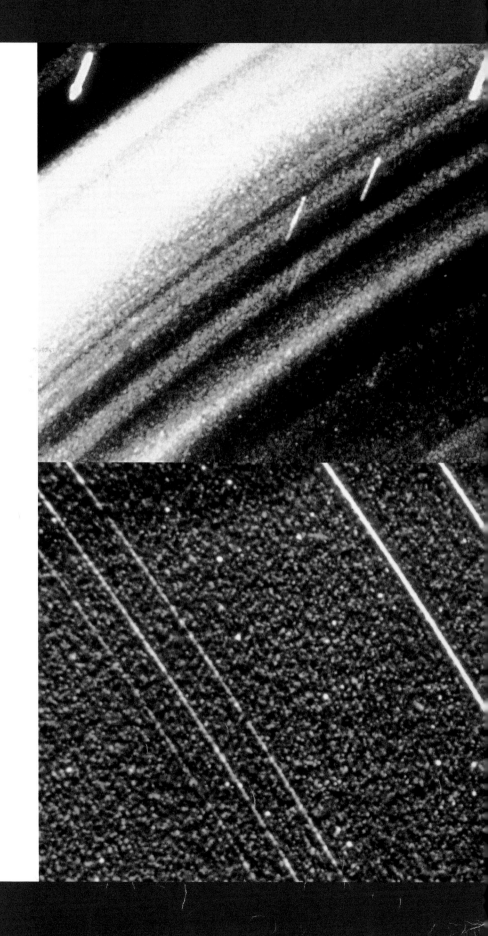

Looking at Uranus from Space

Sometimes astronomers send spacecraft to planets. The spacecraft can look at the planets more closely. *Voyager 2* is the only spacecraft that has visited Uranus. It was launched on August 20, 1977. It flew by Uranus years later on January 24, 1986.

Voyager 2 studied all the gas giants. It was built to visit Jupiter and Saturn. Scientists thought it might last for five years. After visiting Jupiter and Saturn, however, *Voyager 2* was still in great shape. So

◀ *Uranus has eleven icy, rocky rings.*

it flew on to visit Uranus.

Voyager 2 took thousands of pictures of Uranus and its rings and moons. It sent these pictures back to Earth. Finally, scientists learned more about Uranus. Voyager 2 found ten new moons and new rings.

Voyager 2 studied the next planet, Neptune, after Uranus. Then it headed off into space. Now Voyager 2 will explore the solar system beyond the planets. Scientists hope Voyager 2 will last until 2020. That's not too bad for something built to last only five years!

Voyager 2 is prepared for launch.

After studying Jupiter and Saturn, Voyager 2 went on to gather information on Uranus and Neptune.

A Typical Planet

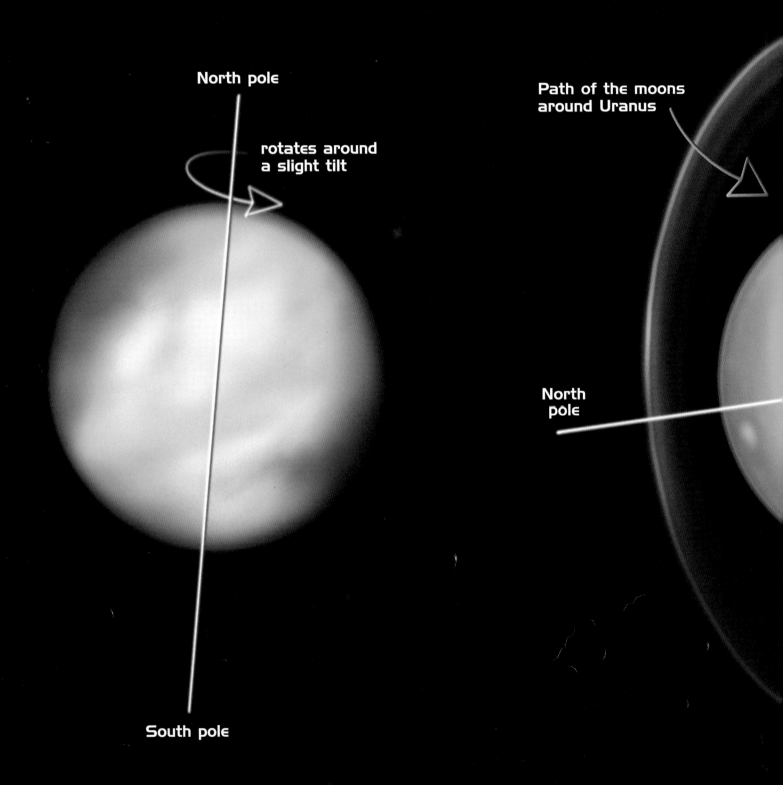

North pole

rotates around a slight tilt

South pole

Path of the moons around Uranus

North pole

Uranus

South pole

rotates on its side

Looking to the Future

✨ *Voyager 2* gave scientists more information about Uranus than they had ever had. But there is always more to discover.

Scientists are still curious about Uranus's tilt. They want to find out what caused it. Also, they are watching Uranus as it goes through its seasons. Uranus's year is very long. It takes scientists a long time to see all there is to see.

No new missions are

◂ *Uranus is a very curious planet to scientists because of its unique tilt.*

planned for Uranus right now. So scientists will be looking closely through their telescopes on Earth. They will also look at pictures from the Hubble Space Telescope. It is hoped they will find exciting answers to their questions about Uranus—the cold and tilted planet.

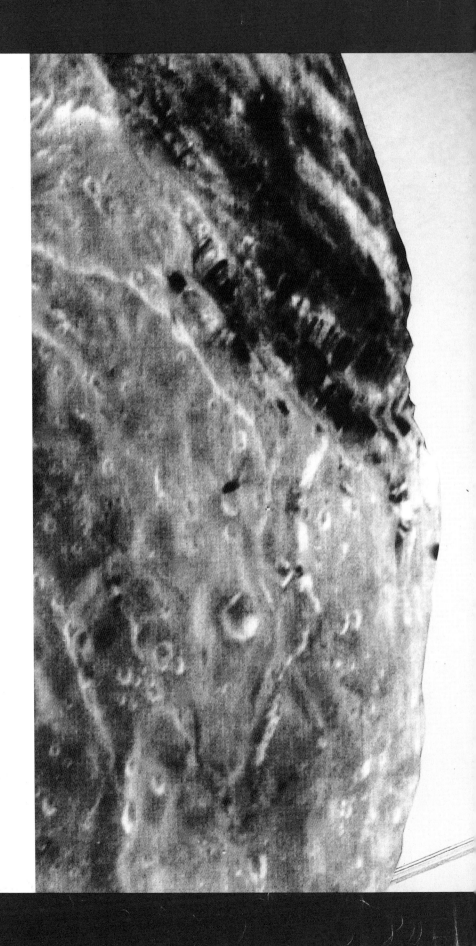

Uranus (opposite page) and its odd moon Miranda (this page) are two of many objects in the solar system that scientists have questions about.

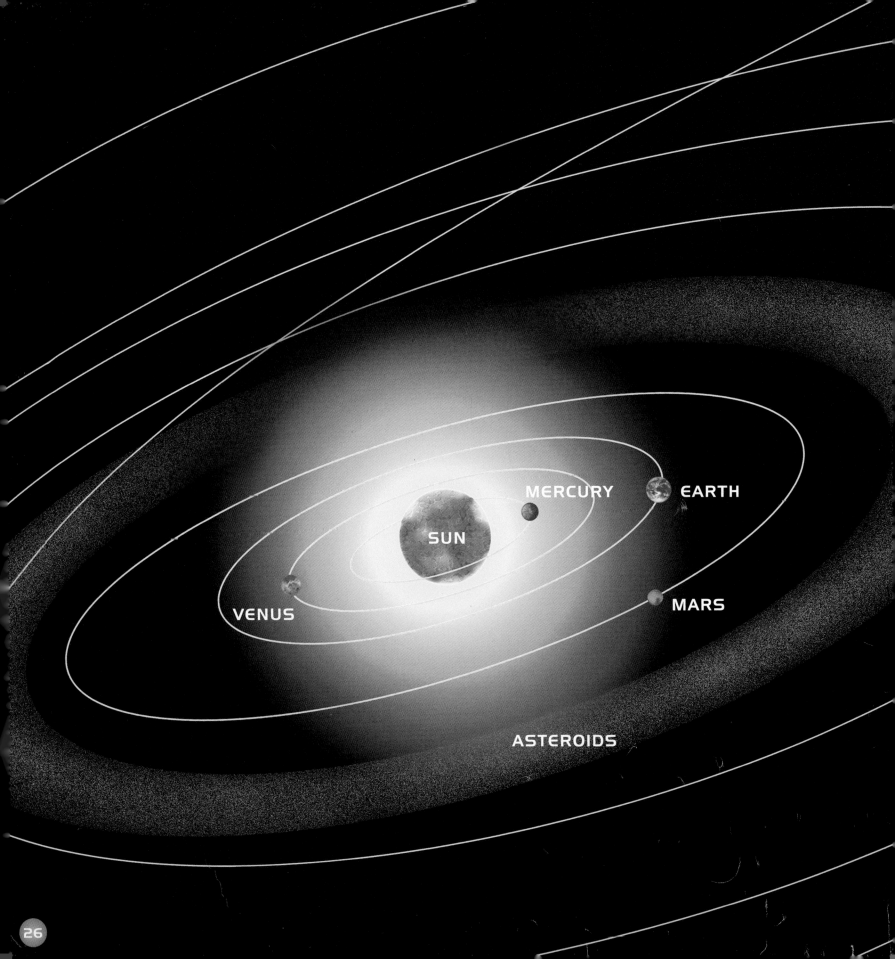

Glossary

astronomer—someone who studies space

binoculars—tools, like a high-powered pair of glasses, that make objects look closer

canyons—deep valleys with steep sides

core—the center of a planet

craters—bowl-shaped landforms created by meteorites crashing into a planet

hurricanes—very windy and strong storms

solar system—a group of objects in space including the Sun, planets, moons, asteroids, comets, and meteoroids

telescope—a tool astronomers use to make objects look closer

temperature—how hot or cold something is

A Uranus Flyby

Uranus is the third-largest planet and the seventh planet from the Sun.

If you weighed 75 pounds (34 kilograms) on Earth, you would weigh 68 pounds (31 kilograms) on Uranus.

Average distance from the Sun: 1,780 million miles (2,870 million kilometers)

Distance from the Earth: 1,605 million miles (2,582 million kilometers) to 1,962 million miles (3,157 million kilometers)

Diameter: 32,000 miles (51,488 kilometers)

Number of times Earth would fit inside Uranus: 63

Did You Know?

- *Voyager 2* was launched at about the same time as *Voyager 1*. A total of five trillion pieces of information about the gas giants were sent back to Earth by the two spacecraft.

- Some people think that the rings around Uranus and its strange tilt make Uranus look like a very large bull's-eye target.

- Each Voyager spacecraft carries a record of sounds and pictures from Earth. Anyone from another planet who found one of the spacecraft would learn all about Earth.

- Two moons discovered by *Voyager 2*, Ophelia and Cordelia, have an interesting job. They keep the rings together. One orbits outside a ring, and the other orbits inside. This keeps the small boulders of the ring in place.

- Uranus's rings are hard to see because the pieces of rock that make them up are almost black.

- Uranus's moons have been named after characters in the plays written by William Shakespeare.

Time it takes to orbit around Sun (one Uranus year): 84 Earth-years

Time it takes to rotate (one Uranus day): 17.25 Earth-hours

Structure: possibly a rocky/iron core surrounded by a slushy or liquid layer

Average temperature: −320° Fahrenheit (−195° Celsius)

Atmosphere: hydrogen, helium, methane

Atmospheric pressure (Earth=1.0): unknown

Moons: 21 known

Rings: 11

Want to Know More?

AT THE LIBRARY
Kerrod, Robin. *Uranus, Neptune, and Pluto*. Minneapolis: Lerner Publications, 2000.
Mitton, Jacqueline, and Simon Mitton. *Scholastic Encyclopedia of Space*. New York: Scholastic Reference, 1998.
Redfern, Martin. *The Kingfisher Young People's Book of Space*. New York: Kingfisher, 1998.
Ridpath, Ian. *Stars and Planets*. New York: DK Publishing, Inc., 1998.
Vogt, Gregory L. *Uranus*. Mankato, Minn.: Bridgestone Books, 2000.

ON THE WEB

Exploring the Planets: Uranus
http://www.nasm.siedu/ceps/etp/uranus/
For more information about Uranus

The Grandest Tour
http://www.jpl.nasa.gov/voyager/
For more information about the Voyager missions

The Nine Planets: Uranus
http://www.seds.org/nineplanets/nineplanets/uranus.html
For a multimedia tour of Uranus

Solar System Exploration: Missions to Uranus
http://sse.jpl.nasa.gov/missions/uranus_missns/uranus-v2.html
For more information about the important NASA mission to Uranus

Space Kids
http://spacekids.hq.nasa.gov
NASA's space-science site designed for kids

Space.com
http://www.space.com
For the latest news about everything to do with space

Star Date Online: Uranus
http://www.stardate.org/resources/ssguide/uranus.html
For an overview of Uranus and hints on where it can be seen in the sky

Welcome to the Planets: Uranus
http://pds.jpl.nasa.gov/planets/choices/uranus1.htm
For pictures and information about Uranus

THROUGH THE MAIL

Goddard Space Flight Center
Code 130, Public Affairs Office
Greenbelt, MD 20771
To learn more about space exploration

Jet Propulsion Laboratory
4800 Oak Grove Drive
Pasadena, CA 91109
To learn more about the spacecraft missions

Lunar and Planetary Institute
3600 Bay Area Boulevard
Houston, TX 77058
To learn more about Uranus and other planets

Space Science Division
NASA Ames Research Center
Moffet Field, CA 94035
To learn more about Uranus and solar system exploration

ON THE ROAD

Adler Planetarium and Astronomy Museum
1300 S. Lake Shore Drive
Chicago, IL 60605-2403
312/922-STAR
To visit the oldest planetarium in the Western Hemisphere

Exploring the Planets* and *Where Next Columbus?
National Air and Space Museum
7th and Independence Avenue, S.W.
Washington, DC 20560
202/357-2700
To learn more about the solar system at this museum exhibit

Rose Center for Earth and Space/Hayden Planetarium
Central Park West at 79th Street
New York, NY 10024-5192
212/769-5100
To visit this new planetarium and learn more about the planets

UCO/Lick Observatory
University of California
Santa Cruz, CA 95064
408/274-5061
To see the telescope that was used to discover the first planets outside of our solar system

Index

Ariel (moon), 16, 17
atmosphere, 12, 14–15
Bode, Johann, 6
canyons, 17
core, 11–12
craters, 16, 17
Elliot, James L., 18
gas giants, 11, 19
George III, king of England, 5
helium, 12
Herschel, Sir William, 5, 16
Hubble Space Telescope, 14, 24
hurricanes, 14–15
hydrogen, 12
Kuiper, Gerard, 16
Lassell, William, 16

methane, 12
Miranda (moon), 16, 17
moons, 16–18
Neptune, 20
Oberon (moon), 16–17
orbit, 8, 18
revolution, 4, 8, 10, 23
rings, 18
rotation, 8, 9
seasons, 4, 10, 23
spacecraft, 11, 18, 19–20
temperature, 15
tilt, 8, 10, 23
Titania (moon), 16, 17
Umbriel (moon), 16, 17
Voyager 2 spacecraft, 19–20, 23

◀ **About the Author:** *Dana Meachen Rau loves to study space. Her office walls are covered with pictures of planets, astronauts, and spacecraft. She also likes to look up at the sky with her telescope and write poems about what she sees. Ms. Rau is the author of more than seventy-five books for children, including nonfiction, biographies, storybooks, and early readers. She lives in Burlington, Connecticut, with her husband, Chris, and children, Charlie and Allison.*